Problem solving
Unit guide
The School Mathematics Project

The right of the
University of Cambridge
to print and sell
all manner of books
was granted by
Henry VIII in 1534.
The University has printed
and published continuously
since 1584.

Cambridge University Press

Cambridge   New York   Port Chester   Melbourne   Sydney

Main authors   Stan Dolan
Tim Everton
Ron Haydock
Tom Patton
Jeff Searle

Team leader   Ron Haydock

Project director   Stan Dolan

Initial work on various case studies was carried out by members of the Spode Group. Many others have helped with advice and criticism.

The School Mathematics Project would like to acknowledge the prior work on mathematical modelling by Mechanics in Action, NE Further Mathematics Scheme and the Open University.

The authors would like to give special thanks to Ann White for her help in producing the trial edition and in preparing this book for publication.

Published by the Press Syndicate of the University of Cambridge
The Pitt Building, Trumpington Street, Cambridge CB2 1RP
40 West 20th Street, New York, NY 10011–4211, USA
10 Stamford Road, Oakleigh, Melbourne 3166, Australia

First published 1991

Produced by Gecko Limited, Bicester, Oxon.

Cover Design by Iguana Creative Design

Printed in Great Britain at the University Press, Cambridge

British Library cataloguing in publication data

16–19 mathematics.
Problem-solving.
Unit guide.
1. Problem-solving
I. School Mathematics Project
515

ISBN 0 521 40808 3

Contents

Introduction to 16–19 Mathematics

Nobody reads introductions and nobody reads teachers' guides, so what chance does the introduction to this Unit Guide have? The least we can do is to keep it short! We hope that you will find the discussion point and tasksheet commentaries and ideas on presentation and enrichment useful.

The School Mathematics Project was founded in 1961 with the purpose of improving the teaching of mathematics in schools by the provision of new course materials. SMP authors are experienced teachers and each new venture is tested by schools in a draft version before publication. Work on *16–19 Mathematics* started in 1986 and the pilot of the course has been used by over 30 schools since 1987.

Since its inception the SMP has always offered an 'after-sales service' for teachers using its materials. If you have any comments on *16–19 Mathematics*, or would like advice on its use, please write to:

16–19 Mathematics
The SMP Office
The University
Southampton SO9 5NH

Why 16–19 Mathematics?

A major problem in mathematics education is how to enable ordinary mortals to comprehend in a few years concepts which geniuses have taken centuries to develop. In theory, our view of how to pass on this body of knowledge effectively and pleasurably has changed considerably; but no great revolution in practice has been seen in sixth-form classrooms generally. We hope that in this course, the change in approach to mathematics teaching embodied in GCSE schemes will be carried forward. The principles applied in the course are appropriate to this aim.

- Students are actively involved in developing mathematical ideas.
- Premature abstraction and over-reliance on algorithms are avoided.
- Wherever possible, problems arise from, or at least relate to, everyday life.
- Appropriate use is made of modern technology such as graphic calculators and microcomputers.
- Misunderstandings are confronted and acted upon.

By applying these principles and presenting material in an attractive way, A level mathematics is made more accessible to students and more meaningful to them as individuals. The *16–19 Mathematics* course is flexible enough to provide for the whole range of students who obtain at least a grade C at GCSE.

Structure of the courses

The A and AS level courses have a core-plus-options structure. Details of the full range of possibilities, including A and AS level *Further Mathematics* courses, may be obtained from the Joint Matriculation Board, Manchester M15 6EU.

For the A level course *Mathematics (Pure with Applications)*, students must study eight core units and a further two optional units. The structure diagram below shows how the units are related to each other. Other optional units are being developed to give students an opportunity to study aspects of mathematics which are appropriate to their personal interests and enthusiasms.

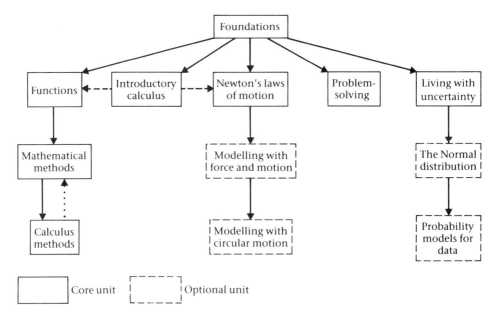

The *Foundations* unit should be started before or at the same time as any other core unit.

Any of the other units can be started at the same time as the *Foundations* unit. The second half of *Functions* requires prior coverage of *Introductory calculus*. *Newton's laws of motion* requires calculus notation which is covered in the initial chapters of *Introductory calculus*.

The Polynomial approximations chapter in *Mathematical methods* requires prior coverage of some sections of *Calculus methods*.

For the AS level *Mathematics (Pure with Applications)* course, students must study *Foundations*, *Introductory calculus* and *Functions*. Students must then study a further two applied units.

Material

In traditional mathematics texts the theory has been written in a didactic manner for passive reading, in the hope that it will be accepted and understood – or, more realistically, that the teacher will supply the necessary motivation and deal with problems of understanding. In marked contrast, *16–19 Mathematics* adopts a questing mode, demanding the active participation of students. The textbooks contain several new devices to aid a more active style of learning.

- Topics are opened up through **group discussion points**, signalled in the text by the symbol

and enclosed in rectangular frames. These consist of pertinent questions to be discussed by students, with guidance and help from the teacher. Commentaries for discussion points are included in this unit guide.

- The text is also punctuated by **thinking points**, having the shape

and again containing questions. These should be dealt with by students without the aid of the teacher. In facing up to the challenge offered by the thinking points it is intended that students will achieve a deeper insight and understanding. A solution within the text confirms or modifies the student's response to each thinking point.

- At appropriate points in the text, students are referred to **tasksheets** which are placed at the end of the relevant chapter. These mostly consist of a self-contained piece of work which is used to investigate a concept prior to any formal exposition. In many cases, it takes up an idea raised in a discussion point, examining it in more detail and preparing the way for formal treatment. There are also **extension tasksheets** (labelled by an E) for higher attaining students which investigate a topic in more depth and **supplementary tasksheets** (labelled by an S) which are intended to help students with a relatively weak background in a particular topic. Commentaries for all the tasksheets are included in this unit guide.

The aim of the **exercises** is to check full understanding of principles and give the student confidence through reinforcement of his or her understanding.

Graphic calculators/microcomputers are used throughout the course. In particular, much use is made of graph plotters. The use of videos and equipment for practical work is also recommended.

As well as the textbooks and unit guides, there is a **Teacher's resource file**. This file contains:

- review sheets which may be used for homework or tests;

- datasheets;

- technology datasheets which give help with using particular calculators or pieces of software;

- a programme of worksheets for more able students which would, in particular, help prepare them for the STEP examination.

Introduction to the unit (for the teacher)

In this unit, students are encouraged to think about the process of mathematical problem-solving rather than the techniques used in this process.

The first four chapters of the text attempt to analyse an extremely complex activity into at least some of its components. Chapter 5 consists of a number of case studies intended to develop skills in the reading and writing of pieces of mathematics. These case studies also serve as sources of exemplars for the processes considered in the first four chapters. The final chapter contains possible starting points for mathematical investigations.

This material will have served its purpose if students, after completing the unit, find themselves both posing their own problems and reflecting much more on the nature of the processes of mathematical activity.

Using this unit

All students should enjoy the challenge of the 'puzzle element' of this unit and should be able to tackle the work at their own level. Discussions arising from the discussion points and exercises work especially well with mixed-ability groups.

- Very able students have plenty to stretch them – providing they are encouraged to take problems further or extend them in various directions. Able students will find the work undemanding only if they are tackling it at too superficial a level.

- Many students welcome the fact that there is often no single 'right answer'! Some students, however, will need the teacher to stress that it is the processes which are being studied and not the mathematics of a particular solution.

Having solutions at the back of the book is of benefit to most students. However, the solutions should be used to check the work rather than as a support. Some teachers have found it helpful to photocopy exercise 1 of chapter 1 before distributing the book itself. As with other 16–19 texts, it is best to make students feel responsible for their own learning strategies and to stress that premature reference to solutions is self-defeating.

One danger of splitting up the problem-solving process into steps is that students may start to apply these steps to problems in a routine manner, irrespective of their appropriateness. This will not occur if exercises, questions and case studies are used frequently as teaching points, with a reflective analysis of the methods adopted. The first chapter opens up with some shortish problems to motivate students and to provide examples which can be used when reflecting upon the processes involved. You may wish to take this further and to adopt an approach based on case studies; first following a problem through and then identifying the stages, rather than taking one problem-solving ingredient at a time, as occurs in the first four chapters.

Some additional notes on the individual chapters may prove helpful.

Chapters 1 and 2

Students do not need to spend inordinate amounts of time solving the problems in the exercises. The problems will have served their purpose if important process ideas have been introduced; ideas which would otherwise be extremely difficult to 'teach'. For example, in exercise 2 of chapter 2 there is no need to try to produce full answers – it is only necessary to pick out important particular cases.

The first discussion point of chapter 2 can be organised as a student activity. Pairs of students can be seated back-to-back with each having to describe a diagram for the other to draw. This is good fun and should bring up many ideas on notation which can be shared later in open discussion.

Chapter 3

Students are likely to be familiar with the idea of a mathematical model in contexts such as 'a model of the British economy', 'a model of North Sea pollution' or 'a model of the traffic flow'. The initial problem on total eclipses has the virtue of being a good example of the processes of making simplifying assumptions and refining a model so that it conforms better to reality. However, students may lack confidence when tackling such an unstructured modelling question. The eclipse problem should therefore be tackled as a class discussion with the teacher using the commentary to structure the discussion. The commentary outlines three mathematical models for this problem. There is no need for the discussion to reach model 3. The refinement of model 1 into model 2 is itself a good illustration of the modelling process. You may, of course, choose to replace the total eclipse problem with a favourite modelling exercise of your own. The problems on the tasksheets can similarly be replaced. What is important is that both teachers and students have problems which seem worthwhile to **them**.

Other units of the 16–19 course, especially the Mechanics units, are sources of modelling exercises. When students have tackled *Living with uncertainty*, it would be possible to discuss probabilistic and deterministic models and, after later Statistics units, you could consider probability models and 'goodness of fit'.

Chapter 4

It should be easy to adapt the text and ideas for the most able students who will be able to take the work of this chapter much further. For example, in tackling sums of series you could develop proof by induction, the method of differences and the use of the results for Σn, Σn^2 and Σn^3.

Chapter 5

It is certainly not necessary to tackle all the case studies. A reasonably large choice is given so that a selection appropriate to the particular class or individual student can be made. You may wish to use one or two of these early on in the unit to illustrate particular processes.

Chapter 6

Coursework gives students the opportunity to exhibit qualities not disclosed by formal examinations. It is hoped that this unit establishes a relevant context for such work. The starting points themselves are only provided as suggestions. A student's own interests may provide a much more fruitful topic for a project.

Tasksheets and resources

1 Mathematical enquiries

1.1 Introduction
1.2 Starting an enquiry
1.3 Particular cases
1.4 Forming patterns
1.5 Generalisation

2 Organising your work

2.1 Notation and symbols
2.2 Classifying
2.3 Tabulation
    Tasksheet 1 or 1E – Tables of differences

3 Mathematical modelling

3.1 Solar eclipses
3.2 Modelling processes
3.3 Modelling exercises
    Tasksheet 1 – Reading age
    Tasksheet 2 – Wallpapering guide
    Tasksheet 3 – Fencing
    Tasksheet 4 – Journeys
    Tasksheet 5 – Sports day (group project)
    Tasksheet 6 – Miscellany

4 Completing investigations

4.1 Counter-example
4.2 Proof
    Tasksheet 1 – Convincing reasons
4.3 Fermat and proof
    Tasksheet 2 – Regions of a circle
    Tasksheet 3E – Prime number formulas
4.4 Extending an investigation
    Tasksheet 4 or 4E – Square and triangular grids

5 Mathematical articles

5.1 Introduction
    Case study 1 – The Platonic solids
    Case study 2 – The gravity model in geography
5.2 Reading mathematics
5.3 Case studies
    Case study 3 – The game of Hex
    Case study 4 – Proportional representation

6 Starting points

1 Mathematical enquiries

1.1 Introduction

You are recommended to spend not more than an hour on each problem in exercise 1, any extra being above and beyond the call of duty. You may need encouragement and possibly the odd hint but otherwise little help.

In answering the problems we hope that you will use, without prompting, some of the methods described later – looking at special cases or analogous simpler problems, guessing patterns, classifying, generalising and so on. Within each problem there is scope for a wide range of sophistication of approach; for instance, in question 1, you might be able to answer (a), (b) and (c) in the case of n competitors and also justify your answers.

We hope that, in the *Problem-solving* unit particularly, you will see mathematics as an activity in which problems are tackled using a variety of techniques – not simply the ones imposed by the textbook or the teacher. You must succeed at your own level if you are to develop the necessary independence of thought to be able to tackle unfamiliar problems with confidence. You must modify, improve and extend your own methods of solution; confidence in your own ability is not increased if you always scrap your own methods in favour of other people's solutions.

Later in the book, various mathematical processes are described. The list is not intended to be exhaustive and a valuable class discussion can take place concerning which other processes should have been included!

1.3 Particular cases

> Which values of n are especially important particular cases in the tournament problem?

The trivial case of $n = 1$ may be mentioned (1 bye and no games). Tournaments with 2^k players (k a positive integer) have no byes. Any cases which lead to insights concerning a problem are especially important.

1.4 **Forming patterns**

Use conjecture and pattern spotting to investigate:

$(1 \times 2 \times 3 \times 4) + 1 =$
$(2 \times 3 \times 4 \times 5) + 1 =$
$(3 \times 4 \times 5 \times 6) + 1 =$

. . .

You probably spotted that the numbers 25, 121 and 361 are squares and took the sequence further using a calculator, obtaining

$$5^2, 11^2, 19^2, 29^2, 41^2, \ldots$$

At this stage, you may wish to introduce the idea of looking at the first differences.

$$5 \top 11 \top 19 \top 29 \top 41$$

first differences   6   8   10   12

This should result in further conjectures, which may be verified.

Question 1 of exercise 3 could now be answered using the pattern established, but for the sake of rigour you will probably want to verify the identity

$$n(n + 1)(n + 2)(n + 3) + 1 \equiv (n^2 + 3n + 1)^2$$

1.5 **Generalisation**

(a) Express s_3 as the sum of two triangular numbers.

(b) Express s_4 as the sum of two triangular numbers.

(c) Express s_n as the sum of two triangular numbers.

This discussion point provides an opportunity to check that you understand subscript notation.

(a) $s_3 = t_3 + t_2$ (b) $s_4 = t_4 + t_3$ (c) $s_n = t_n + t_{n-1}$

At this stage, it is sufficient to give a geometrical justification of the generalisation.

$$t_3 + t_2 = s_3$$

2 Organising your work

2.1 Notation and symbols

> How good are you at describing objects or giving clear directions without using a street plan?
>
> Suppose you have to describe a diagram, such as the one below, over the telephone. How could you do it?

Labelling the points is a useful first step.

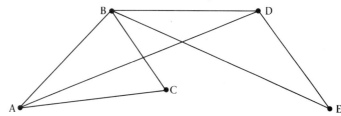

You would need to say which pairs of points are linked, for example A–B, A–C, etc. You might also give distances and directions between points or, alternatively, you might give the approximate coordinates of each point.

2.2 Classifying

> Name at least three different classes of
>
> (a) quadrilateral,   (b) function.

You may be able to recall cases of the use of classification from elsewhere in the course, say the use of acute-angled, right-angled and obtuse-angled triangles in establishing the cosine rule. The discussion point provides an opportunity to discuss not only the classes but also the relationships between them.

(a) parallelograms, quadrilaterals having exactly one line of symmetry, rectangles, etc.;

(b) polynomial, exponential, self-inverse, etc.

3

Tables of differences

1 As an example, you might take

$$E(n) = n^2 - n + 2$$

Then $E(1) = 2$, $E(2) = 4$, $E(3) = 8$, etc.

$$2 \downarrow 4 \downarrow 8 \downarrow 14 \downarrow 22$$
$$2 \downarrow 4 \downarrow 6 \downarrow 8$$
$$2 \quad 2 \quad 2$$

Whichever quadratic expression was chosen the second differences should be equal.

You may have realised that there was an unnecessary restriction in this question. *a, b* and *c* can be any numbers but, of course, it simplifies calculations if they are integers.

2 The third differences should be equal.

3

Sequence		-1	-6	3	68	255	654	1379	. . .
First differences			-5	9	65	187	399	725	. . .
Second differences				14	56	122	212	326	. . .
Third differences					42	66	90	114	. . .
Fourth differences						24	24	24	. . .

On the basis of the pattern established in questions 1 and 2, you should conjecture that the sequence was generated by a quartic expression.

Tables of differences

1 As an example, you might take

$$E(n) = n^2 - n + 2$$

Then $E(1) = 2$, $E(2) = 4$, $E(3) = 8$, etc.

$$
\begin{array}{ccccccccc}
2 & & 4 & & 8 & & 14 & & 22 \\
& 2 & & 4 & & 6 & & 8 & \\
& & 2 & & 2 & & 2 & &
\end{array}
$$

Whichever quadratic expression was chosen the second differences should be equal.

You may have realised that there was an unnecessary restriction in this question. a, b and c can be any numbers but, of course, it simplifies calculations if they are integers.

The first differences are in arithmetic progression and second differences are equal.

2 The third differences should be equal.

3 (a) First differences are all 1.

(b) Second differences are all 2.

(c) Third differences are all 6.

(d) Fourth differences are all 24 (see below).

Sequence (n^4)	1		16		81		256		625		1296		2401	...
First differences		15		65		175		369		671		1105		...	
Second differences			50		110		194		302		434		...		
Third differences				60		84		108		132		...			
Fourth differences					24		24		24		...				

4 From the answers to question 3, a generalisation is that n^k generates a sequence giving kth differences all $k!$ (and hence $(k+1)$th differences all 0). So a polynomial with leading term n^5 would generate a sequence with fifth differences all $5! = 120$. The likely answer to the question is that the sequence was generated by a polynomial of fifth degree, leading term $\frac{1}{4}n^5$.

3 Mathematical modelling

3.1 Solar eclipses

How long does a total eclipse last?

The main dimensions needed are (from *Encyclopedia Britannica*)

diameters in km:	sun	1.39×10^6
	earth	1.28×10^4
	moon	3.48×10^3
mean distances from earth in km:	sun	1.50×10^8
	moon	3.82×10^5

The moon orbits the earth in 27.3 days, a lunar month.

FIRST MODEL

In a simple model, you might consider the sun and earth to be stationary and the moon's orbit around the earth to be circular. You can define an eclipse as total when, for some time, the moon completely obscures the main disc of the sun from some observer O on earth.

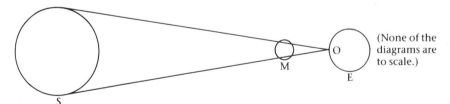

(None of the diagrams are to scale.)

A partial eclipse begins with the moon in position M_1 and ends in position M_4. The total eclipse is between positions M_2 and M_3.

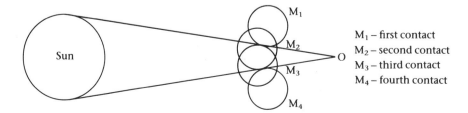

M_1 – first contact
M_2 – second contact
M_3 – third contact
M_4 – fourth contact

To make further progress, you can simplify your model by assuming the moon's orbit is locally straight.

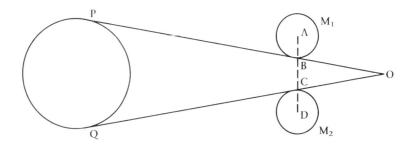

You may also make the assumptions that PQ is the sun's diameter and BC is part of the moon's orbit.

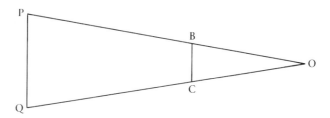

Triangle OPQ is an enlargement of triangle OBC with centre O, scale factor $\dfrac{OP}{OB}$ and so

$$BC = PQ \times \frac{OB}{OP} = 1.39 \times 10^6 \times \frac{3.82 \times 10^5}{1.50 \times 10^8}$$

$$= 3.54 \times 10^3 \, km$$

The diameter of the moon is 3.48×10^3 km so, according to the model, total eclipses do not occur! Since $3.48 < 3.54$, the moon never completely obscures the sun. So you must refine the model in some way.

SECOND MODEL

One simplification you made was that the orbits of the earth around the sun and of the moon around the earth are circular. In fact, both are roughly elliptical; instead of using mean distances, you should consider a range of distances.

	moon	sun
nearest distance from earth (km)	3.63×10^5	1.47×10^8
furthest distance from earth (km)	4.06×10^5	1.52×10^8

The eclipse of maximum duration will occur when the moon is nearest the earth and the sun furthest away. Applying the model in this case,

$$BC = 1.39 \times 10^6 \times \frac{3.63 \times 10^5}{1.52 \times 10^8} = 3.32 \times 10^3 \, \text{km}$$

The eclipse is total only when BC falls within the diameter of the moon, for a distance of

$$(3.48 - 3.32) \times 10^3 = 1.6 \times 10^2 \, \text{km}$$

The moon's speed is

$$\frac{\text{circumference of total circular orbit}}{\text{lunar month}} = \frac{2\pi \times 3.82 \times 10^5}{27.3 \times 24 \times 60}$$

$$= 61.1 \, \text{km per minute}$$

(The elliptical orbit has a length about equal to that of a circle with radius the mean distance.)

Using the relation

$$\text{time} = \frac{\text{distance}}{\text{speed}}$$

$$\text{time of total eclipse} = \frac{1.6 \times 10^2}{61.1}$$

$$= 2.6 \, \text{minutes}$$

The maximum duration of total eclipse is observed to be about 7 minutes. Can your model be amended further to account for this?

THIRD MODEL

Another main assumption made at the outset was that the earth could be regarded as stationary. In fact the observer O is moving because of the spin of the earth on its axis. The speed of an observer is greatest on the equator, being

$$\frac{\text{circumference of earth}}{\text{day length}} = \frac{\pi \times 1.28 \times 10^4}{24 \times 60}$$

$$= 27.9 \, \text{km per minute}$$

Since the observer and the moon move in the same direction, the speed of the moon relative to the observer is

$$\text{moon's speed in orbit} - \text{observer's speed} = 61.1 - 27.9$$

$$= 33.2 \, \text{km per minute}$$

Using this figure, you will find that the time of total eclipse is

$$\frac{1.6 \times 10^2}{33.2} = 4.8 \text{ minutes}$$

You still have not reached the target and a dutiful modeller would check assumptions and consider other refinements.

You should at least consider the following relevant observations.

(1)   Over a series of calculations, a considerable rounding error accumulates.

(2)   The distances between bodies are probably 'centre-to-centre'; this is not always made clear in reference books.

(3)   The moon's speed in elliptical orbit is not constant.

(4)   Orbits are not quite elliptical anyway! In particular, the sun perturbs the orbit of the moon.

(5)   In observations not made from the equator, you must take into account the latitude in calculating the speed of the observer.

Since 1963, scientists have used specially equipped aircraft or observation platforms above the clouds. Supersonic machines flying at 2000 km per hour in the direction of the moon's shadow prolong the period of total eclipse to an hour or more.

3.2 Modelling processes

> Consider your solution to the total eclipse problem. Try to identify the four stages of the modelling process described above.

It is important that you should reflect upon **your** solution. As an illustration of the modelling process, the main text contains an outline of how the four stages can be identified in a solution to the total eclipse problem.

Reading age

As indicated on the tasksheet, your method will depend on the type of book you are studying. The following commentary assumes you are considering a method for assessing the reading age of books for young children.

You might consider a particular reading scheme and gather data on various variables according to the stated age range of material in the scheme.

Age	Number of words of one syllable	Average number of letters per word	etc.
4			
5			
6			
etc.			

SET UP MODEL/ANALYSE

You would then try to fit the data with a reasonably simple formula. Some of your chosen variables may not appear to be relevant and can be discarded.

INTERPRET/VALIDATE

You might then consider other schemes and check if your method gives a reading age in agreement with these schemes. If it does not, then you must decide if it is your method which is not appropriate or the scheme itself! Discussions with primary school teachers should prove valuable.

Finally, you can apply your (possibly revised) method to reading books not in a scheme and decide its range of applicability.

Wallpapering guide

As in all real problems, reference should be made as far as possible to actuality. For example, in answering question 1, you should take measurements from your own room and actual rolls of paper. There is no single right approach.

1 (a) The problem is about covering and so you should list

- length, width and height of room,
- dimensions of doors and windows,
- length and breadth of a roll of paper.

A less obvious dimension needed is the design repeat length of the chosen paper. In practice, whether you get three or four full lengths out of a roll is crucial in determining the number of rolls needed. If you only get three pieces then a simple strategy is

count the number of full length pieces needed, and divide by 3.

(Any shorter lengths needed over and under windows, over doors etc. may be obtained from the offcuts.)

Other measurements might include those of soffits and returns (the horizontal and vertical areas around doors, windows and any alcoves).

(b) Length and width and any returns might be added to give a notional perimeter; smaller windows and other features might be ignored. Perhaps it would be defeating the object of the exercise to enquire whether the shop would take back whole unused rolls!

(c) Because of the importance of the number of full-length pieces obtainable from a roll, one approach is indicated by the incomplete flow chart:

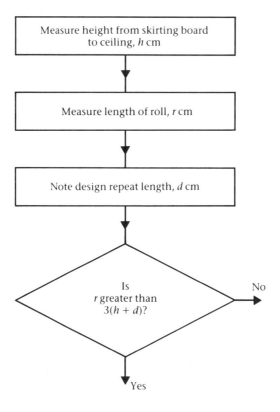

Measure height from skirting board to ceiling, h cm

Measure length of roll, r cm

Note design repeat length, d cm

Is
r greater than
$3(h + d)$?

No

Yes

and the strategy to be used will involve branching.

On the other hand, you might use a method based on areas. A simple first attempt might be to divide the total wall area (including doors and windows), A, by the area of a roll of paper, S, giving a simple formula $\dfrac{A}{S}$. The complications introduced by the need to match the pattern bring in a 'matching factor' k, so that $\dfrac{kA}{S}$ rolls are needed. From the previous discussion it appears that an upper bound for k is $\frac{4}{3}$.

A further factor should be introduced to account for the fact that no horizontal joins will be allowed. You could call this the 'patching factor'.

Finally, you might like to compare an 'area' and a 'counting' model.

Fencing

1 The simplest model would allow for 21 posts, 200 m of railing and, say, 2 man-days of labour. Estimates for each of these should be given.

2 Further details might include:

- wastage of timber,
- transport of material and labour,
- the possible need for stronger posts at the ends of fencing and some intermediate points,
- nails and other minor items (the option of including the treatment of the fence with preservative could be mentioned),
- VAT,
- the contractor's profit.

The model could include a breakdown of the timing of the work based on the number of saw cuts and the time taken for each, the average time to sink a post, and so on. A more refined estimate could then be made of the labour costs.

3,4 The model can be generalised by using letters instead of figures. The cost of a particular fence can then be obtained by substituting appropriate figures. Some skill must be exercised in achieving a simple formula or table which can be used easily by the person responsible for making the estimate.

Your model can be validated through discussion with a fencing contractor. It should be interesting to discover what particular 'rules of thumb' they use.

Journeys

You may wish to impose more structure, possibly using a local journey you know well, so that you can check times and fuel consumption against reality. However, finding a 'right answer' is less important than the clear communication of ideas in a written report.

Probably the best way to deal with the tasksheet is to spend a short time at the end of a lesson discussing the problem in class, finding maps and other necessary information at home, possibly also writing the report as homework.

A simple model would include:

- distances obtained from a motoring atlas;
- estimated average speeds on motorways, dual carriageways, urban and other roads.

Refinements might include:

- taking account of rush-hour congestion by estimating waiting times at intersections according to local conditions;
- considering fuel consumptions at various speeds for the chosen type of car (a consideration which gives plenty of scope for mathematical modelling in its own right!).

Sports day

The PE department can probably help with advice and recommendations on distances and dimensions.

Constraints to be used in setting up the initial model might include these:

- At least two events may occur simultaneously.
- Adequate safety margins must be allowed around throwing areas.
- The 100 m track must be straight.
- The radius of curvature on other tracks must not be too small.
- Suitable arrangements must be made for spectators and all events must be visible to them.

When drawing up a programme of events and working out how it could be run on the field, you will need to refine your model. Not all the constraints may be possible and some priority among them will have to be decided. You may find it necessary to allow spectators to move about to see all the events. You should consider arrangements for starters, judges and recorders.

Miscellany

1   (a) Road fund tax and insurance are certainly relevant. Depreciation in the value of
        the car should be included. The remaining costs depend on the extent to which
        the car is used. Variables include distances travelled, rates of petrol
        consumption, cost per litre of fuel, and number and cost of services. Costs of
        other repairs, of garaging and of membership of a motoring organisation are
        also possibilities.

    (b) It could be assumed either that current prices will continue to apply or that
        some or all of the prices might be multiplied by a factor related to the projected
        increase in the cost of living. Rates of petrol consumption for different kinds of
        motoring (urban, motorway, etc.) might be lumped together to give an average.

    (c) The type of formula obtained, consisting of a fixed term and a variable one
        dependent on distance travelled, is met in many modelling problems.

        Remember that you are trying to find a **real** solution and not just an abstract
        formula. The AA regularly publishes figures for car running costs of the
        following type:

        Running a car: annual standing and running costs (pence per mile)

	Up to 1000 cc	1001–1400 cc	etc.
5000 miles	38	48	
10 000 miles	25	30	
etc.			

        Find such a table of figures and compare it with your result.

2   For a simple model you might assume that the molecules from Julius Caesar's
    breath still exist as gas molecules and are randomly dispersed throughout the
    atmosphere.

    You must then estimate the total number of gas molecules in a breath, B, and the
    total number in the atmosphere, A. A simple analysis shows the expected number
    you inhale with each breath to be $\dfrac{B^2}{A}$.

    A deep investigation would need to consider the nature of the gases exhaled by
    Julius Caesar and the likelihood of, for example, a carbon dioxide molecule having
    been used in photosynthesis.

TASKSHEET
COMMENTARY **6**

3 (a) You would certainly require the current costs of first and second class postage and also the proportion of letters sent at each rate.

(b) For n_1 first class stamps costing c_1 pence and n_2 second class stamps costing c_2 pence,

$$n_1 c_1 + n_2 c_2 \leq 100$$

(c) The design will depend upon the current values of the variables above. Including stamps of slightly greater value than the charge for the book and making up the amount with a low-value stamp are two devices which have been used by the Post Office.

4 Completing investigations

4.2 Proof

> When you have a conjecture of your own, how can you be sure that no-one will be able to find a counter-example?

The discussion can be tailored to fit both your interest and the time available. There is an opportunity here, for example, to recall proofs met previously, such as the deductive proofs of geometry and the proof by contradiction that $\sqrt{2}$ is irrational (*Foundations*) or you might prefer to consider the difference between a scientific law, which is strengthened by the accumulation of evidence but not normally proved (for example Newton's inverse square law of gravitation: a mathematical model later refined by Einstein), and a mathematical 'law', which must remain merely a conjecture until proved.

Other results which should lead to lively discussion are

$$\sin^2 x + \cos^2 x = 1$$

and the result that

the angle subtended by a chord at the centre of a circle is twice the angle it subtends at the circumference.

4.4 **Extending an investigation**

> (a) Show how the answers to the three problems are analogous.
>
> (b) In the first problem, the five points need not be in line. Use this idea to deduce the number of diagonals of a pentagon.

(a) The number of line segments is $\binom{5}{2}$, but to bring out the analogy this is written as

$$1 + 2 + 3 + 4$$

Then the number of squares is

$$1^2 + 2^2 + 3^2 + 4^2$$

and the number of cubes is

$$1^3 + 2^3 + 3^3 + 4^3$$

(b) Of the ten line segments, five are sides of the pentagon. The remaining five are diagonals.

Convincing reasons

1 (a) $2n$ is $2 \times n$ and is therefore even because it is divisible by 2.

   (b) $2n + 1$. This number is $2 \times n + 1$ and is therefore odd because it has a remainder of 1 when divided by 2.

   (c) $(2n + 1) + (2n + 3) + (2n + 5) = 6n + 9 = 3 \times (2n + 3)$
   is divisible by 3.

2 (a) $10b + a$

   (b) $11a + 11b = 11 \times (a + b)$

   (c) The rule does **not** work for three-digit numbers, for example $102 + 201 = 303$.
   The rule **does** work for four-digit numbers.
   $(1000a + 100b + 10c + d) + (1000d + 100c + 10b + a)$
   $$= 1001a + 110b + 110c + 1001d$$
   $$= 11 \times (91a + 10b + 10c + 91d)$$

3 (a) $(10a + b) \times 11 = 110a + 11b$
   $$= 100a + 10a + 10b + b$$
   $$= 100a + 10(a + b) + b$$
   So 'ab' $\times 11 \quad = $ '$a\ a{+}b\ b$'

Regions of a circle

1

n	1	2	3	4
r_n	1	2	4	8

n is the number of points, and r_n is the corresponding number of regions.

Apparently the number of regions doubles every time a point is added, i.e.

$r_{n+1} = 2 \times r_n$, in which case

$r_n = 2^{n-1}$

2 $r_5 = 16$, which agrees with the conjecture. However, $r_6 = 31$, so the conjecture is false.

3

n	1	2	3	4	5	6	7
r_n	1	2	4	8	16	31	57

4

First differences		1		2		4		8		15		26
Second differences			1		2		4		7		11	
Third differences				1		2		3		4		
Fourth differences					1		1		1			

5 The formula for r_n is probably a quartic polynomial.

The polynomial

$r_n = an^4 + bn^3 + cn^2 + dn + e$

can be found by solving five equations (for the five unknown coefficients), obtained by substituting corresponding values of n and r_n. The first is

$a + b + c + d + e = 1$

and the second is

$16a + 8b + 4c + 2d + e = 2$

The study could be taken further, either as a class or by an individual student. See *Mathematics in School* 2.2, March 1973.

Prime number formulas

3E

It is worth spending some time in exploring particular cases and establishing subsidiary results such as that if $an^2 + bn + c$ is always to be odd then c must be odd and a and b must have the same parity.

In the first attempt at a proof, the cases $c = \pm 1$ and $ac + b + 1 = \pm 1$ all invalidate the method. It is worth remembering that any one of these suffices as a counter-example.

When $n = 1 + 2p$,

$$an^2 + bn + c = a(1 + 2p)^2 + b(1 + 2p) + c$$
$$= 4ap^2 + (4a + 2b)p + a + b + c$$
$$= (4ap + 4a + 2b + 1)p$$

The second attempt is sound and the method may be extended to show that no polynomial form $P(n)$ of degree one or more generates only primes. If the degree of the polynomial is k it is sufficient to show that $P(n)$ takes the same value for $(k + 1)$ different values of n.

Square and triangular grids

1 (a) 16   (b) 9   (c) 4   (d) $1 + 4 + 9 + 16 = 30$

2 This time there will be

           25 1×1 squares
           16 2×2
             9 3×3
and   4 4×4 squares.

The total is $1 + 4 + 9 + 16 + 25 = 55$.

3 There are

           16 triangles of edge 1,
            7 triangles of edge 2,
and   3 triangles of edge 3.

The total is $1 + 3 + 7 + 16 = 27$.

4 Using a similar method, the total is $1 + 3 + 6 + 13 + 25 = 48$.

Square and triangular grids

1 4; namely $1 \times 1, 2 \times 2, 3 \times 3$ and 4×4 squares.

2 There are

   16 1×1 squares
   9   2×2 squares
   4   3×3 squares.

So the total is $1 + 4 + 9 + 16 = 30$.

3 Using a similar method of counting, the total number is

   $1 + 4 + 9 + \ldots + n^2 = \frac{1}{6}n(n + 1)(2n + 1)$

4 There are 27 in all, including 16 of edge 1, 7 of edge 2 and 3 of edge 3.

5 (a) 13   (b) 48

6 Classifying into triangles pointing upwards ($\triangle$) and those pointing downwards ($\triangledown$) and making separate counts for the two classes, you will find patterns occurring in which sequences of triangular numbers appear, for example in the $4 \times 4 \times 4$ grid there are

   $1 + 3 + 6 + 10$ pointing upwards

and

   $1 + 6$ pointing downwards.

For the $6 \times 6 \times 6$ grid, using the patterns found, there should be

   $1 + 3 + 6 + 10 + 15 + 21$ upwards

and

   $1 + 6 + 15$ downwards,

giving a total of 78 triangles.

5 Mathematical articles

5.3 Case studies

 What reading strategy should you adopt when studying a piece of mathematical writing?

For many mathematical articles three readings are advisable:

- to note the gist and the main conclusions;
- to study the arguments used to reach those conclusions;
- to check details of arguments for rigour and calculations for accuracy.

If a piece is particularly difficult it may be advisable to 'sleep on it' or discuss it with a fellow student or teacher.

The Platonic solids

> (a) What do you understand by the term 'regular polygon'?
>
> (b) What is meant by 'infinitely many' regular polygons?

(a) A 'polygon' is a closed plane figure bounded by straight edges.

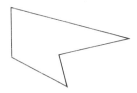

A 'regular polygon' is a polygon all of whose edges are equal and all of whose angles are equal.

(b) There is no limit to the number of different regular polygons that can be drawn.

    . . .

> Name a three-dimensional shape which fits the description of a Platonic solid.

As will be seen, there are actually five Platonic solids. In view of the picture in the text, the **cube** is the one most likely to be chosen to answer this question.

> Explain fully why six triangular faces cannot meet at a vertex.

The triangular faces must all be equilateral triangles and so the angles at the vertex from the six faces would total precisely 360°. The six triangles would then each be part of a single flat hexagonal face and would not in fact be individual faces.

What are the values of F, V and E for a cube?

$F = 6$, $V = 8$, $E = 12$

Find $F + V - E$ for (a) the octahedron and (b) the cube.

(a) $F + V - E = 8 + 6 - 12 = 2$ (octahedron)

(b) $F + V - E = 6 + 8 - 12 = 2$ (cube)

Use Euler's formula to show that a Platonic solid with three triangular faces at each vertex must have precisely four faces.

The shape is formed by F triangular faces with a total of $3F$ sides and $3F$ corners. As before,

$$E = \frac{3F}{2}$$

Each vertex of the solid is at a corner of each of three faces and so

$$V = \frac{3F}{3}$$

Then

$$V + F = E + 2$$

$$\Rightarrow F + F = \frac{3F}{2} + 2$$

$$\Rightarrow \quad F = 4$$

Therefore

$$E = \frac{3F}{2} = 6, \quad V = F = 4$$

The solid is a tetrahedron.

> Explain the assertion that three square faces must meet at each vertex.

As before, there must be at least three faces at each vertex. Four or more faces would contribute an angle count of at least $4 \times 90°$, in which case the faces would not form a corner. There are therefore precisely three faces at each vertex.

> Explain why $E = \dfrac{4F}{2}$ and $V = \dfrac{4F}{3}$.

There are F square faces, having $4F$ sides and $4F$ corners. When joined together, each edge is formed from a pair of sides, so

$$E = \frac{4F}{2}$$

and each vertex is formed from three corners, so

$$V = \frac{4F}{3}$$

> What is the significance of the 108°?

Each angle of a regular pentagon is 108°.

> Obtain the given values of F, V and E.

$$E = \frac{5F}{2} \text{ and } V = \frac{5F}{3}$$

From Euler's formula,

$$F + \frac{5F}{3} = \frac{5F}{2} + 2$$

Multiply both sides by 6,

$$6F + 10F = 15F + 12 \quad \Rightarrow \quad F = 12$$

Then $\quad V = \dfrac{5F}{3} = 20 \quad$ and $\quad E = \dfrac{5F}{2} = 30$

The gravity model in geography

The motivation for an inverse square model may be worth examining, especially if Newton's law is not familiar. You might compare the intensity of gravitational attraction with the intensity of light.

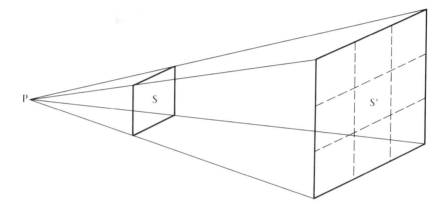

Think of a cine projector P illuminating in turn two screens S and S′. If S is three times as far from P as S then the area of light on S′ is nine times that on S. But the same amount of light falls on both S and S′. So the intensity of light on S′ is only $\frac{1}{9}$ of that on S. In fact, the intensity is inversely proportional to the square of the distance from P. (Replace 'cine projector' by 'candle' to give the analogy the right Newtonian flavour!)

> What value does β have in Newton's model of gravitation?

$\beta = 2$

> (a) Use the model to find T_{DN} and T_{SN}.
>
> (b) Which pair of the towns Sheffield, Derby and Nottingham has the most telephone calls between them according to the model?

(a) $T_{\text{DN}} = \dfrac{k \times 216 \times 271}{20^2} \approx 146k, \qquad T_{\text{SN}} = \dfrac{k \times 537 \times 271}{50^2} \approx 58k$

(b) Derby and Nottingham

29

What is the attractiveness of B to the individual?

$$\frac{kP_B}{d_B{}^2}$$

Explain the calculation that leads to $a = 0.043$.

$$WK = 130a + 173a = 303a$$

so in mile units,

$$303a = 13$$
$$a \approx 0.043$$

(a) Explain the statement

$$\frac{x + 13}{x} = \frac{173}{130}$$

(b) Shrewsbury and Welshpool are 19 miles apart and have populations 56 000 and 7 000 respectively. Using the same gravity model, write down expressions for the attractiveness of the two towns to an individual who lives distance d_S from Shrewsbury and d_W from Welshpool.

(c) Using the method outlined above, carry out the necessary calculations and describe as fully as possible Welshpool's area of influence.

(d) In areas of the USA, it is easy to travel long distances by car. What difference would this make to the value of β taken in the model?

(a) Y divides WK externally in the ratio $130 : 173$, i.e.

$$\frac{WY}{YK} = \frac{130}{173}$$

Since $WY = x$ and $YK = (x + 13)$

$$\frac{x}{x + 13} = \frac{130}{173}$$

(b) Expressions for the attractiveness of Shrewsbury and Welshpool are respectively

$$\frac{56\,000k}{d_S^2} \quad \text{and} \quad \frac{7\,000k}{d_W^2}.$$

(c) The expressions for attractiveness are equal when

$$7d_S^2 = 56d_W^2$$
$$d_S^2 = 8d_W^2$$
$$d_S \approx 2.83d_W$$

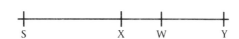

Welshpool's area of influence is the circle diameter XY where X, Y divide SW internally and externally in the ratio $2.83 : 1$. In mile units,

$$SW = 19, \quad XW = \frac{1}{3.83} \times 19 = 4.96, \quad \frac{SY}{YW} = 2.83$$

so if $YW = x$,

$$\frac{x + 19}{x} = 2.83$$

$$x = \frac{19}{1.83} \approx 10.38$$

$$XY = 10.38 + 4.96 \approx 15.3$$

Welshpool's area of influence is a circle of diameter 15.3 miles, centred at a point 2.7 miles from Welshpool on the side opposite Shrewsbury.

(d) The 'distance decay' would be less marked, making β smaller.

The value $\beta = 1$ is often used in US models.

The game of Hex

1   If Black plays at 7, White plays at 5, and connects 1–2/4–5–8/9.

   If Black plays at 3 or 6, White plays at 4 and connects 1–4–7/8.

2   (a)

   (b)   The claim is false. Black connects 9/13–10–11/14–15–12/16.

3   *If White plays at 7*
   White 7 can be connected to the top left edge by 7–3/4. It can be connected to the bottom right edge by Result 1. White therefore cannot be stopped from completing a winning chain.

   If White plays at 10
   The fact that White cannot be prevented from completing a winning chain follows from the result above by symmetry (rotate the board 180°).

Proportional representation

1   Example 1 illustrates the principle of proportional representation as simply as possible. It is unrealistic because votes in exact multiples of 2000 are most unlikely; the figures have been arranged so that proportional representation works out exactly.

2   Party C has three times as many votes as party A but only twice the number of seats. If C were awarded three seats then there would be a total of six seats but there are only five to allocate. Alternatively, removing a seat from one of the other parties would seem unfair on them.

3   (a)

	Party A	Party B	Party C	Party D
	17 920	11 490	11 170	4420
divide by 2	8960	5745	5585	2210
divide by 3	5973	3830	3723	1473
divide by 4	4480	2873	2793	1105

The five highest numbers are: party A – 17 920
party B – 11 490
party C – 11 170
party A –   8960
party A –   5973

The electoral quotient is 5973 giving:
party A – 3.0;   party B – 1.9;   party C – 1.9;   party D – 0.7.

The election result is:
party A – 3 seats;   party B – 1 seat;   party C – 1 seat;   party D – 0 seats.

(b)   Party A has the majority of the seats (3 out of 5) but only obtained 40% of the total votes cast.

On the basis of this there seems to be an advantage in the d'Hondt rule to the party with the largest share of the vote.

Magic squares

1   The given square is 'more up-to-date' because it incorporates the date 1989. The numbers would not be consecutive because $89 - 19 = 70$ and there are only 16 cells in all.

2

⁻1	⁻6	12	⁻1
8	3	⁻5	⁻2
⁻10	3	3	8
7	4	⁻6	⁻1

3   If the given square represents the linear combination

$$aB_1 + bB_2 + cB_3 + dB_4 + eB_5 + fB_6 + gB_7$$

then you obtain the equations

$$a + b = 3, \quad f = 6, \quad e + g = 12, \quad c + d = 7, \quad c + e = 8, \quad d + g = 11,$$
$$a + f = 7, \quad b = 2, \quad d + f = 10$$

At this stage there are enough equations to find all the required coefficients. They are

$$a = 1, \quad b = 2, \quad c = 3, \quad d = 4, \quad e = 5, \quad f = 6, \quad g = 7$$

4   The best strategy is probably to build up the square and the corresponding linear combination of members of G simultaneously, following this procedure:

(i)   fill in the date of birth;

(ii) allocate arbitrary coefficients to B_6, B_2, B_3 and B_7;

(iii) calculate the coefficients of B_1 and B_4 (using the date of birth);

(iv) calculate the coefficient of B_5 (using, say, the magic sum in the fourth row).

5 In the linear combination

$$aB_1 + bB_2 + cB_3 + dB_4 + eB_5 + fB_6$$

the first entry in the fourth row is 0, whereas in B_7 it is 1.

Similarly, B_2, B_3 and B_6 cannot be expressed as linear combinations of the remaining Bs.

Could you write

$$B_1 = pB_2 + qB_3 + rB_4 + sB_5 + tB_6 + uB_7$$

instead? This again yields a contradiction: comparing elements in the first row, second column, $t = 0$; but comparing elements in the second row, third column, $t = 1$.

Similarly, B_4 and B_5 cannot be expressed as linear combinations of the remaining Bs.

So, in any given linear combination of the members of G no substitution is possible for any one of the members to enable the formation of an equivalent linear combination of some subset of G. You can deduce that G is an irreducible generating set for 4×4 magic squares of the type described.

6 If D is the Dürer square, suppose you had an alternative way of expressing D as a linear combination of the members of G. Then, say,

$$D = 8B_1 + 8B_2 + 7B_3 + 6B_4 - 2B_5 + 3B_6 + 4B_7$$

$$= aB_1 + bB_2 + cB_3 + dB_4 + eB_5 + fB_6 + gB_7$$

where corresponding coefficients are not all equal. As an example, suppose $a \neq 8$ (any other non-equal coefficients would serve equally well, in case $a = 8$). Then

$$(8 - a)B_1 = (b - 8)B_2 + (c - 7)B_3 + (d - 6)B_4 + (e + 2)B_5 + (f - 3)B_6 + (g - 4)B_7$$

$$B_1 = \frac{b - 8}{8 - a} B_2 + \frac{c - 7}{8 - a} B_3 + \ldots + \frac{g - 4}{8 - a} B_7$$

which is an expression of B_1 as a linear combination of the remaining Bs, contrary to the result of question 5.

You can deduce that the given linear combination constituting D is **unique**. Similarly, any other magic square is expressible **uniquely** as a linear combination of the members of G.

Communications satellites

1 The centre of the earth must be the centre of the orbit **and** the spin of the earth must result in the satellite maintaining its position over a fixed point on earth. To achieve both of these requirements, the orbit must be above the equator.

2 The period of the earth's spin on its axis is 24 hours.

3 The length of the orbit is $2\pi R$ and the speed of the satellite is v, so the time for a complete orbit is $\dfrac{2\pi R}{v}$. A complete orbit takes one day or $24 \times 60 \times 60$ seconds.

4

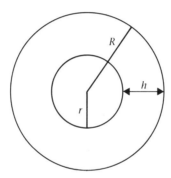

The height of the satellite h is given by $(R - r)$, where r is the radius of the earth.

$$R \approx 4.22 \times 10^7 \,\text{m}$$
$$r \approx 6.4 \times 10^6 \,\text{m} = 0.64 \times 10^7 \,\text{m}$$

Hence

$$h \approx (4.22 - 0.64) \times 10^7 \,\text{m} \approx 3.6 \times 10^7 \,\text{m} = 3.6 \times 10^4 \,\text{km}$$

5 In his second law, Newton, in effect, equated force and mass $\times$ acceleration. The force is $\dfrac{GMm}{R^2}$, the mass is m, so the expression for acceleration is $\dfrac{v^2}{R}$.

CASE STUDY
COMMENTARY **6**

6 The equations corresponding to (1) and (2) become

$$R = \frac{10 \times 60 \times 60}{\pi} v \qquad (1)$$

and

$$Rv^2 = (6.67 \times 10^{-11}) \times (8 \times 10^{23}) \qquad (2)$$

Dividing (2) by (1), you obtain

$$v^3 = \frac{6.67 \times 8 \times 10^{12} \times \pi}{10 \times 60 \times 60}$$

$$v \approx 1.67 \times 10^3$$

Substituting back into (1),

$$R \approx 1.91 \times 10^7$$

The velocity is $1.67 \times 10^3 \, \text{m s}^{-1}$ and the height above the planet's surface is

$$(1.91 \times 10^7 - 0.3 \times 10^7) \, \text{m} = 1.61 \times 10^4 \, \text{km}$$

Queues on the M5

1 All cars are assumed to be of the same length to keep the model simple and manageable.

2 1 mile $\approx \dfrac{8}{5}$ km = 1600 m

3 You are told that N cars pass the observer every minute and that N cars pass the observer in $\dfrac{9ND}{4V}$ seconds. So $\dfrac{9ND}{4V}$ seconds is one minute, or 60 seconds.

$$\frac{9ND}{4V} = 60$$

$$9ND = 240\,V$$

$$N = \frac{240V}{9D} = \frac{80V}{3D}$$

4 The value of $\dfrac{V}{3} + \dfrac{V^2}{60}$

(i) when $V = 30$ is $\dfrac{30}{3} + \dfrac{900}{60} = 10 + 15 \qquad = 25,$

(ii) when $V = 50$ is $\dfrac{50}{3} + \dfrac{2500}{60} = 16.7 + 41.7 = 58.4,$

(iii) when $V = 70$ is $\dfrac{70}{3} + \dfrac{4900}{60} = 23.3 + 81.7 = 105.$

These values correspond closely to the total stopping distances given, the term $\dfrac{V}{3}$ giving the thinking distance and the term $\dfrac{V^2}{60}$ the braking distance.

5   The overall stopping distance is that required between the front bumper of a car and the rear bumper of the preceding car.

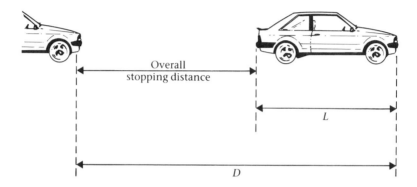

6   (a)   From the table, N is a maximum when the speed is 15 m.p.h.

(b)   At this speed, the separation between cars is

$$\left(\frac{15}{3} + \frac{225}{60}\right)\,\text{m} = 8.75\,\text{m} = \frac{8.75}{4}\,\text{car lengths}$$

or 2.2 car lengths approximately.

Stock control

1 n is the number of orders per year and t_1 is the time after which the first order is made. Suppose t_1 is measured in years and that orders are made every 3 months. Then you should have $n = 4$ and $t_1 = \dfrac{1}{4}$. In general

$$t_1 = \frac{1}{n}$$

Since orders are made at equal intervals of time, $t_2 = 2t_1$, $t_3 = 3t_1$ and so on ($t_k = kt_1$).

2   The demand is the quantity ordered per year, $D = nS$.

3

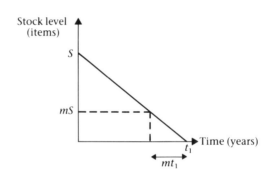

Stock is assumed to be sold at a uniform rate, so that the stock level is proportional to the time remaining before the next order. As a result, the graph of S against t is a straight line in the interval $0 \le t < t_1$.

4   With mixed stock (more than one kind of item) it may be possible to stimulate demand by displaying a large amount of stock, so that orders might be larger than an EOQ based on previous sales. In such a case the criterion would probably be to maximise profit or sales. In another case, storage space might be limited to a number less than the EOQ. A little thought will suggest other possibilities.

5   (a) $EOQ = \sqrt{\dfrac{200Dp}{Vr}} = \sqrt{\dfrac{200 \times 300 \times 35}{200 \times 12}} = 29.6$ (say 30)

(b) $n = \dfrac{D}{S} = \dfrac{300}{30} = 10.$

This indicates that 10 orders should be placed per year.

CASE STUDY

COMMENTARY **8**

(c) $C = \dfrac{Dp}{S} + \dfrac{SVr}{200} = \dfrac{300 \times 35}{30} + \dfrac{30 \times 200 \times 12}{200} = 350 + 360 = 710$

The annual cost is £710.

(d) If V increases to 240, no other change being made, then the new EOQ is

$$\sqrt{\frac{200 \times 300 \times 35}{240 \times 12}} = 27$$

This represents a 10% decrease on the old order quantity.

6 (a) $\dfrac{\mathrm{d}C}{\mathrm{d}S} = \dfrac{-Dp}{S^2} + \dfrac{Vr}{200} = 0$

$$\Rightarrow \quad S^2 = \frac{200\,Dp}{Vr}$$

$$\Rightarrow \quad S = \sqrt{\frac{200\,Dp}{Vr}}$$

7 If $x < 20$, $\quad C = \dfrac{26 \times 12 \times 15}{x} + \dfrac{x \times 24 \times 18}{200}$

$$C = \frac{468}{x} + 2.16x$$

If $20 \leq x < 50$, $\quad C = \dfrac{468}{x} + 1.80x$

If $x \geq 50$, $\quad C = \dfrac{468}{x} + 1.44x$

A table of values for the function $x \rightarrow C$ is given below.

	x	5	10	15	20	25	30	35	40	45	50	55	60	65	70
($V = 24$)	C	104.4	68.4	63.6	(66.6)										
($V = 20$)					59.4	63.7	69.6	76.4	83.7	91.4	(99.4)				
($V = 16$)											81.4	87.7	94.2	100.8	107.5

From the following graph, you can see that the EOQ is 20 (when the cost is £59.40).

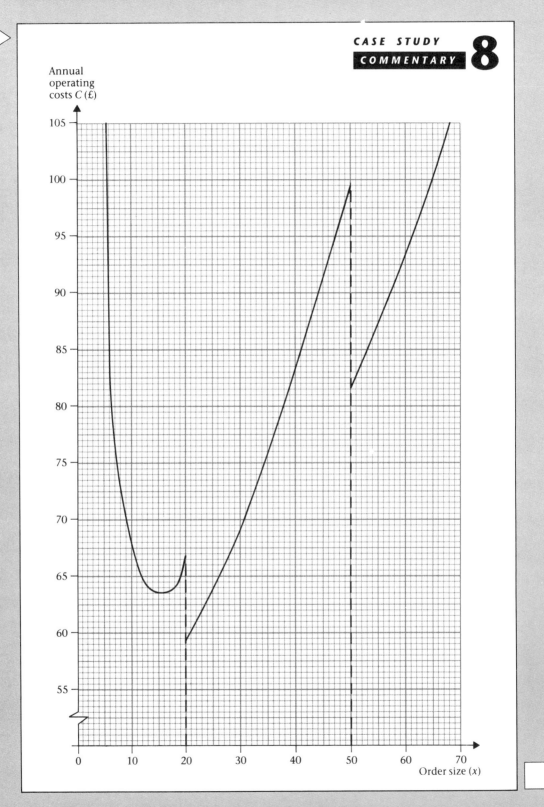

Annual operating costs C (£)

Order size (x)

Pecking orders

1 The case with $n = 2$, when A pecks B or B pecks A, is trivial.

2 The relation 'pecks' is represented by an arrow, thus: A ➤ B, meaning A pecks B.

3 The two diagrams both illustrate a transitive relation. If A and B were interchanged in the second diagram it would represent precisely the same situation as the first. Networks are said to be 'of the same type' when they show the same number of vertices linked in the same way, irrespective of the labelling of the vertices, the lengths of the lines, and so on.

4 The network does not show a transitive relation because in the triangle BCD, the arrows proceed cyclically. In a transitive relation, if B pecks C and C pecks D, then B must peck D.

5 In the product $n(n - 1)$ all the corners have been taken in turn and for each corner, all the lines meeting there have been counted. Since every line joins two corners, each line is counted twice by this method; for example the line AB is counted among the lines meeting at A and also among those meeting at B.

6

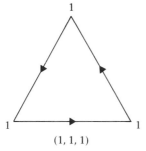

(1, 1, 1)   (2, 1, 0)

7

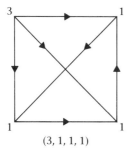   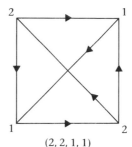

(3, 1, 1, 1)   (2, 2, 1, 1)

8 (a) The numbers in a dominance set count the arrows proceeding away from each corner in turn. Since every line has just one arrow, the sum of all the numbers is equal to the number of lines.

(b) In a transitive network with n vertices, the dominance set is

$$\{(n-1),(n-2),\ldots,2,1,0\}$$

Thus, in the transitive network with four vertices illustrated

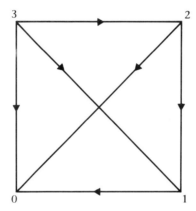

the dominance set is (3, 2, 1, 0).

Archimedes and π

1 An upper bound for the circumference is a value above which the circumference cannot possibly lie. Similarly, a lower bound is a value below which the circumference cannot possibly lie.

The actual value lies between the upper and lower bounds. The actual value is known with greater precision as the lower and upper bounds are brought closer together – precision improves as the bounds converge.

2 (a)

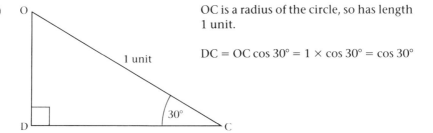

OC is a radius of the circle, so has length 1 unit.

$DC = OC \cos 30° = 1 \times \cos 30° = \cos 30°$

(b)

PR is a tangent to the circle, with point of contact at S. OS is a radius of the circle and so has length 1 unit.

$SR = OS \cot 30° = 1 \times \cot 30° = \cot 30°$

3 Perimeter of $\triangle ABC = 6 \times DC = 6 \cos 30° = 3\sqrt{3}$

Perimeter of $\triangle PQR = 6 \times SR = 6 \cot 30° = 6\sqrt{3}$

Thus $3\sqrt{3} <$ circumference $< 6\sqrt{3}$

or $3\sqrt{3} < 2\pi < 6\sqrt{3}$

or $\dfrac{3\sqrt{3}}{2} < \pi < 3\sqrt{3}$

4 (a) An inscribed figure is a figure that is totally within the circle but touches the circle at its vertices.

(b) A circumscribed figure is a figure that is totally outside the circle but such that its edges form tangents to the circle.

5 M is the point of contact of the tangent PQ with the circle, so OM is a radius and has length 1 unit. OA and OB are also radii so have length 1 unit. $\triangle$OAB is equilateral since ABCDEF is a regular hexagon, i.e. OA = OB = AB. Hence AB has length 1 unit. The perimeter of ABCDEF is $6 \times AB = 6$ units.

6 (a) In $\triangle$OLB, OB = 1 unit, since it is a radius.
LB = $\frac{1}{2}$ unit, since L is the mid-point of AB.

Angle OLB = 90°, so by Pythagoras' theorem, OL = $\sqrt{1 - \left(\frac{1}{2}\right)^2} = \frac{\sqrt{3}}{2}$.

So OM = 1 and OL = $\frac{\sqrt{3}}{2}$.

The scale factor of the enlargement is

$$\frac{OM}{OL} = \frac{1}{\left(\frac{\sqrt{3}}{2}\right)} = \frac{2}{\sqrt{3}}.$$

Hence PQ = $\frac{2}{\sqrt{3}} \times AB = \frac{2}{\sqrt{3}}$

and the perimeter of PQRSTU = $6 \times PQ = \frac{12}{\sqrt{3}} = \frac{4 \times 3}{\sqrt{3}} = 4\sqrt{3}$.

(b) Perimeter of ABCDEF < circumference < perimeter of PQRSTU

$6 < 2\pi < 4\sqrt{3}$
$3 < \pi < 2\sqrt{3}$

7

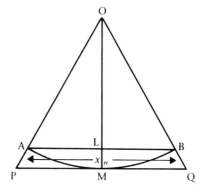

Let AB be a side of the regular inscribed n-gon so that

$$AB = x_n.$$

Then $LB = \dfrac{x_n}{2}$, since L is the mid-point of AB.

OB has length 1 unit.

By Pythagoras' theorem in $\triangle$ OLB,

$$OL = \sqrt{1^2 - (\tfrac{1}{2}x_n)^2} = \sqrt{1^2 - \tfrac{1}{4}x_n^2}$$

The scale factor of the enlargement is

$$\frac{OM}{OL} = \frac{1}{\sqrt{1^2 - \tfrac{1}{4}x_n^2}},$$

so $PQ = \dfrac{AB}{\sqrt{1 - \tfrac{1}{4}x_n^2}} = \dfrac{x_n}{\sqrt{1 - \tfrac{1}{4}x_n^2}}$.

Multiplying by the number of sides n, in order to obtain the perimeter, gives

$$C_n = \frac{I_n}{\sqrt{1 - \tfrac{1}{4}x_n^2}}$$

8 (a) $x_{12} = \sqrt{\tfrac{1}{4}x_6^2 + [1 - \sqrt{1 - \tfrac{1}{4}x_6^2}]^2}$

(b) $x_6 = 1$ since x_6 is the length of the side of the inscribed hexagon.

$$x_{12} = \sqrt{\tfrac{1}{4} + [1 - \sqrt{1 - \tfrac{1}{4}}]^2} \approx 0.518$$

9 $I_{12} = 12x_{12} = 12 \times 0.518 = 6.212$

$$C_{12} = \frac{I_{12}}{\sqrt{1 - \frac{1}{4}x_{12}{}^2}} = 6.658$$

n	x_n	I_n	C_n	Inequality
12	0.518	6.212	6.658	$3.106 < \pi < 3.329$

(All calculations are quoted to 3 decimal places, having been evaluated using $x_{12} = 0.517\,638$.)

10 As the number of sides of a regular n-gon increases, the n-gon becomes a closer and closer approximation to the circle itself. So the inscribed and circumscribed n-gons both have perimeters that approach the circumference of the circle, getting arbitrarily close as n increases without limit. By increasing n, increasingly accurate approximations to π are obtained. The difference between the approximation and π itself can be made arbitrarily small by taking a sufficiently large value of n.

6 Starting points

6.2 'Pure' problems

It is important that 'solutions' should **not** be provided for problems in this chapter! Some hints are indicated for just two of the problems.

1 BILLIARDS

(i) What would be the smallest table?

(ii) In what simple way can a path be described 'mathematically'? Are there any other ways?

(iii) Look at several particular easy cases before attempting to generalise.

2 CHAINS – DIVISORS

You do not need to investigate all these suggestions. A perfectly satisfactory project might consider just one in depth. Other equally acceptable ideas for investigation might occur to you.

6.3 'Real' problems

It is worth spending a little time in discussing a strategy. For many problems, a possible procedure is as follows.

(i) List relevant variables and minimise the list using any relationships. Make reasonable assumptions.

(ii) Set up the model, analyse and interpret.

(iii) Check its validity using appropriate sources and methods (books of reference, leaflets and newspapers, by enquiries in the field, etc.).

(iv) Discuss the assumptions and the model generally with others and particularly those with relevant knowledge.

(v) Amend the model and repeat the 'modelling loop' as necessary.

(vi) Write a report, which will normally include some account of rejected models and the reason for their rejection.

In problems such as 7 (Magazine sales) and queueing problems (at traffic lights, petrol pumps, cash points, check-outs, etc.), a probabilistic model will be appropriate. It may be possible to analyse the model using a simulation exercise, with dice or a spinner providing the appropriate chance element. Physical models may be useful in solving other problems. You should adopt an adventurous and creative approach.